Probability

by Sarah Jane Brian

SCHOLASTIC
PROFESSIONAL BOOKS

New York ✳ Toronto ✳ London ✳ Auckland ✳ Sydney

Edited by Sarah Glasscock

Cover design by Jaime Lucero and Vincent Ceci

Interior design by Ellen Matlach Hassell
for Boultinghouse & Boultinghouse, Inc.

Interior illustrations by Kate Flanagan and Manuel Rivera

ISBN 0-590-37367-6

Contents

Probability of a Simple Event

Probability as Fraction, Decimal, Percent, and Ratio

Sample Space

Counting Outcomes

TREE DIAGRAMS AND FUNDAMENTAL COUNTING PRINCIPLE

PERMUTATIONS AND COMBINATIONS

COUNTING OUTCOMES VISUALLY

(continued on the next page)

* Activity includes a student reproducible.

Making Predictions

Probability of a Compound Event

Real-World Applications

✳ Activity includes a student reproducible.

Introduction

With this book of activities, part of a six-book mathematics series, we hope to make teaching and understanding probability fun, creative, and exciting.

An Overview of the Book

Table of Contents

The table of contents features the activity names and page numbers, as well as stars to mark student reproducibles. Activities are categorized by probability topic, so you may use the table of contents as a scope and sequence.

Teaching Pages

Everything you need to know is on the teaching page, but you also have the option of tailoring the activities to meet students' individual needs and to address the wide variety of skills displayed in your classroom.

Learning Logo

A logo indicating the probability topic being discussed appears at the top of the page. The logo is correlated to the topics in the table of contents. This will enable you to key the activities to your mathematics curriculum quickly and easily.

Learning Objective

The objective clearly states the primary aim of the activity.

Grouping

This states whether the whole class, individual students, pairs, or cooperative groups should perform the task. If an activity lends itself to more than one grouping, the choices are indicated. Again, if you feel that another grouping is more appropriate to your classroom, feel free to alter the activity accordingly.

Materials

To cut your preparation time, all materials necessary for the main activity (including student reproducible) and its extension are listed. Most of the materials are probably already in your classroom. If an activity has a student reproducible with it, the page number of the reproducible is listed here.

Advance Preparation

A few activities require some minimal advance preparation on your part. All the directions you need are given here. You may also let students take over some or all of the preparation.

Directions

The directions usually begin with suggestions on how to introduce or review the probability topic, including any terms and/or formulas. Step-by-step details on how to do the activity follow. When pertinent, specific strategies that might help students in solving problems are suggested.

Taking It Farther

This section on the teaching page offers suggestions on how you can extend and enrich the activity. Students who require extra help and those who need a challenge will both benefit when you move the activity to a different level.

Assessing Skills

The key questions and/or common errors pointed out in this section will help alert you to students' progress. (In fact, you may want to jot down more questions on the page.) Use the information you gather about students here in conjunction with the teacher assessment form that appears on page 64 of the book.

Answers

When answers are called for, they appear at the bottom of the teaching page.

Student Reproducibles

About one-third of the activities have a companion student reproducible page for you to duplicate and distribute. These activities are marked with a star in the table of contents.

Do I Have Problems!

These pages are filled with fun and challenging Problems of the Day that you may write on the board or post on the bulletin board. The answers appear in brackets at the end of each problem.

Assessment

Student Self-Evaluation Form

At the end of the activity, hand out these forms for students to complete. Emphasize that their responses are for themselves as well as you. Evaluating their own performances will help students clarify their thinking and understand more about their reasoning.

Teacher Assessment Form and Scoring Rubric

The sign of a student's success with an activity is more than a correct answer. As the NCTM stresses, problem solving, communication, reasoning, and connections are equally important in the mathematical process. How a student arrives at the answer—the strategies she or he uses or discards, for instance—can be as important as the answer itself. This assessment form and scoring rubric will help you determine the full range of students' mastery of skills.

National Council of Teachers of Mathematics Standards

The activities in this book, and the rest of the series, have been written with the National Council of Teachers of Mathematics (NCTM) Standards in mind. The first four standards—Mathematics as Problem Solving, Mathematics as Communication, Mathematics as Reasoning, and Mathematical Connections—form the philosophical underpinning of the activities.

Standard 1: Mathematics as Problem Solving
The open-ended structure of the activities, and their extension, builds and strengthens students' problem-solving skills.

Standard 2: Mathematics as Communication
Class discussion at the beginning and ending of the activities is an integral part of these activities.

Additionally, communication is fostered when students work in pairs or cooperative groups and when individuals share and compare work.

Standard 3: Mathematics as Reasoning
Communicating their processes in working these activities gives students the opportunity to understand and appreciate their own thinking.

Standard 4: Mathematical Connections
A variety of situations has been incorporated into the activities to give students a broad base on which to apply mathematics. Topics range from real-life experiences (historical and contemporary) to the whimsical and fantastic, so students can expand their mathematical thinking to include other subject areas.

More specifically, the activities in this book address the following NCTM Standards.

NCTM Standards Grades K–4:

Standard 11: Statistics and Probability

✳ Collect, organize, and describe data.

✳ Construct, read, and interpret displays of data.

✳ Formulate and solve problems that involve collecting and analyzing data.

✳ Explore concepts of chance.

NCTM Standards Grades 5–8:

Standard 11: Probability

✳ Model situations by devising and carrying out experiments or simulations to determine probabilities.

✳ Model situations by constructing a sample space to determine probabilities.

✳ Appreciate the power of using a probability model by comparing experimental results with mathematical expectations.

✳ Make predictions that are based on experimental or theoretical probabilities.

✳ Develop an appreciation for the pervasive use of probability in the real world.

Toy Joy

Toy Joy will hit students when they design a spinner game to win their favorite toys.

➜ Directions

1. Duplicate a copy of the *Toy Joy* reproducible for each student.

2. Draw a replica of the spinner on the board. With colored chalk, color two of the spinner sections red and two of the spinner sections yellow. Ask students: *If you spin the spinner, which color would you most likely spin? Why?* [red or yellow because each color takes up an equal amount of the spinner] Discuss students' responses.

3. Distribute the reproducible. Tell students that they will be designing a spinner game. The object of the game is to spin to win a toy. They begin by writing down their favorite toy, and then three other toys. Ask students to create a spinner that will give them the most likely chance of spinning their favorite toys. They must use at least two different toys and two different colors in their spinners.

4. Show students how to cut out and assemble their spinners. Let students spin their spinners 10 times and record their results. Discuss the results. Did they design a winning spinner?

★ Taking It Farther

Have students draw spinners with five, six, or eight sections and use them in the experiment. How do their strategies and results change?

✔ Assessing Skills

Observe whether students understand that using fewer toys in their spinners increases the likelihood of winning their favorite toys.

LEARNING OBJECTIVE

Students explore the meaning of probability.

GROUPING

Individual

MATERIALS

✷ *Toy Joy* reproducible (p. 9)
✷ colored chalk
✷ crayons
✷ scissors
✷ paper clip and pencil (to make the spinner)

ANSWERS

The spinner with the most likely chance would have three sections for the favorite toy and one section for another toy.

Toy Joy

What is the likelihood of winning your favorite toy? Design a winning spinner and find out!

Write down your favorite toy on a sheet of paper. Then write down three other toys. Use at least two toys in the spinner below. Write a toy in each section. Color the sections. Cut out your spinner and spin it 10 times. Record your results.

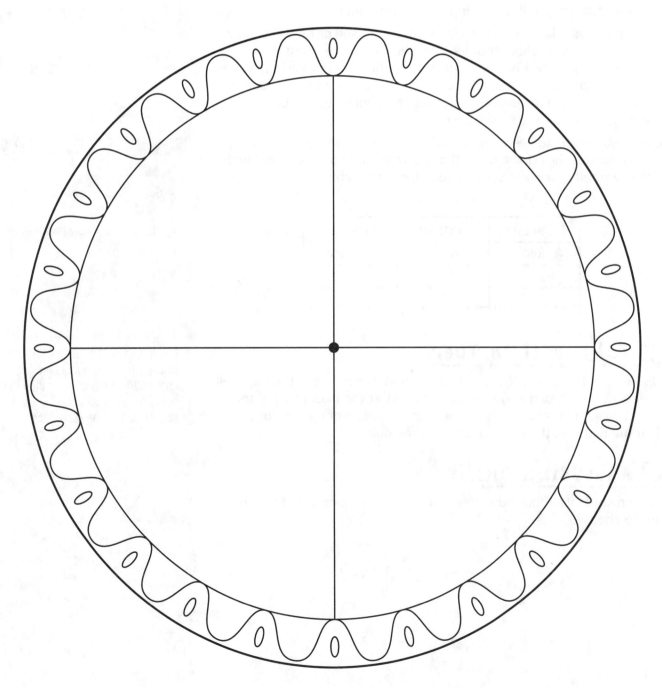

Guess 'n' Go

In this probability game, students guess which color counter they will draw.

🔄➤ Directions

1. Pair students and then pass out the game bags and reproducibles. Explain that they will be playing a guessing game.

2. The partners take turns. Each guesses whether she or he will draw a red counter or a yellow counter from the bag. If the guess is incorrect, the player moves 1 space. If the guess is correct, the player moves 3 spaces for a red counter and 2 spaces for a yellow counter. Players set counters aside—do not return counters to the bag. Whoever reaches the finish space first is the winner.

3. Remind players to keep a record of their guesses and the colors of the counters they draw. Draw the following table on the board with the example. Have each player copy the blank table.

I Guessed	I Drew	I Moved
Red	Yellow	1 space

☆ Taking It Farther

At the end of the game, there will be counters remaining in the bag. Tell students that each bag contained 20 counters at the beginning of the game. Challenge them to guess how many counters of each color are still in the bag. Ask them to discuss their reasoning.

✔ Assessing Skills

Question students about how they made their decisions about which color to choose.

LEARNING OBJECTIVE

Students try to predict the number of red and yellow counters inside a bag.

GROUPING

Pairs

MATERIALS

For each pair:

* *Guess 'n' Go* reproducible (p. 11)
* 20 red and yellow counters
* paper bag
* game markers

ADVANCE PREPARATION

Prepare bags of counters for each pair. Each bag should contain a total of 20 counters. Vary the ratio of red to yellow counters in each bag. Duplicate the reproducible.

Guess 'n' Go

This game is in the bag—all you have to do is guess and draw the right color counter!

Decide which player will go first. Each player guesses whether he or she will draw a red or a yellow counter out of the bag. Here's how you can move on the game board:

An incorrect guess . . . . Move 1 space.

A correct guess (Yellow) . . . . Move 2 spaces.

A correct guess (Red) . . . . . . Move 3 spaces.

Flip for Probability!

**How many times will the coin come up heads? Your class
will flip over this chance to test how probability works.**

➤ Directions

1. Write the following definition on the board:

   Probability of event = P(event) = $\dfrac{\text{Number of Favorable Outcomes}}{\text{Number of Possible Outcomes}}$

2. Discuss the definition of probability with the class. Talk about the
   terms *event, experiment,* and *outcome.* In probability, an *experiment* is
   a trial, such as flipping a coin. The *outcome* is the result.

3. In the experiment of flipping a coin, one outcome is heads. Ask:
   What is the probability of getting heads? The answer is:

   P(heads) = $\dfrac{1 \text{ (Number of Favorable Outcomes: heads)}}{2 \text{ (Number of Possible Outcomes: heads or tails)}}$

   Explain that this can be expressed as an equivalent decimal or
   percent: 0.5, or 50%.

4. Then ask: *If you flip a coin 10 times, how many times would you predict
   that you would get heads?* [$\frac{1}{2}$ of 10, or 5 times] Direct each group to
   flip a coin 10 times and record how many times heads comes up.

5. Have students predict how many times they will get heads if they flip
   the coin 30 times. [$\frac{1}{2}$ of 30, or 15 times] Each group conducts the
   experiment, flipping the coin and recording the results.

6. Let students repeat the experiment, this time flipping 100 times.

7. Allow time for each group to share their results with the class. What
   trends do they notice in the results? [Answers will vary; however, the
   general trend should be that the more times the coin is flipped, the
   closer the outcome gets to the prediction.]

✪ Taking It Farther

Ask students: *What if you flipped two coins at once? What is the probability
of both coins landing on heads?* [$\frac{1}{4}$, or 25%; the possible outcomes for
the two coins are: (heads, heads), (heads, tails), (tails, heads), (tails,
tails).] Have students repeat the experiments, this time flipping two
coins to see if (heads, heads) comes up $\frac{1}{4}$ of the time.

✔ Assessing Skills

To determine if students understand the ratio at work in probability,
note whether they are able to convert the fractions from their
experiments to the equivalent percents.

LEARNING OBJECTIVE

Students find the probability
of a simple event and use an
experiment to test their
conclusions.

GROUPING

Cooperative groups

MATERIALS

✴ pencil and paper
✴ coins

Odds and Evens

Are these games fair or unfair? Students use probability to decide and then play the games to see if they're right.

⟿ Directions

1. Remind students than an *event* is the outcome of something that happens. For instance, in rolling a number cube, rolling a specific number is an event. Review the ratio that is used to express the probability of an event:

   Probability of event = P(event) = $\dfrac{\text{Number of Favorable Outcomes}}{\text{Number of Possible Outcomes}}$

2. Divide the class into pairs and distribute the reproducible.

3. Partners work together to complete the table in the reproducible, answer the questions, and then test their answers by playing the game. The table may be completed by the entire class if students are unsure of how to complete it on their own.

4. Have pairs play the game again, but instead of adding the two numbers, they multiply. Compare the results of the two games. The outcomes are the same. Tailor questions 2a–6 on the worksheet to fit the new game. [**2a.** 27 **b.** $\frac{27}{36}$, or $\frac{3}{4}$, or 75% **3a.** 9 **b.** $\frac{9}{36}$, or $\frac{1}{4}$, or 25%. **4–6.** Answers will vary. The game is unfair because there are many more ways to get an even number than to get an odd number.]

★ Taking It Farther

Ask students to make up games or think of existing games where the probability of winning can be calculated. For example, suppose you must pick an ace from a deck of cards. The probability is $\frac{4}{52}$, or $\frac{1}{13}$. Have them test each game many times to see if the predicted probability works. (Note: In the case of picking from a deck of cards, the card must be returned to the deck before the next card is picked to keep the same probability of picking an ace.)

✓ Assessing Skills

Observe whether students understand the importance of the order of the number cubes. For example, throwing (3, 4) is a separate outcome from (4, 3). Thus, there are two ways to throw a combination of a 3 and a 4.

LEARNING OBJECTIVE

Students find the probability of simple events and experiment to test their conclusions.

GROUPING

Pairs

MATERIALS

For each pair:

* *Odds and Evens* reproducible (p. 14)
* 2 number cubes
* pencil and scrap paper

ANSWERS

1. Row 1—3,1; 6,1; Row 2—3,2; 5,2; Row 3—2,3; 4,3; 6,3; Row 4—3,4; 4,4; 6,4; Row 5—2,5; 4,5; 5,5; 6,5; Row 6—3,6; 5,6

2a. 18 **b.** $\frac{18}{36}$, or $\frac{1}{2}$, or 50%

3a. 18 **b.** $\frac{18}{36}$, or $\frac{1}{2}$, or 50%

4–6. Answers will vary. The game is fair because there is an equal number of ways to get an even or odd number.

Odds and Evens

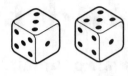

What are your chances of winning this game?
Probability will help you find out!

**Toss two number cubes 25 times. Each time the two numbers
add up to an even number, Player 1 gets one point. Each time
they add up to an odd number, Player 2 gets one point.
Whoever has more points after 25 tosses wins.**

BEFORE YOU PLAY:

1. List the possible
   outcomes for each
   toss in the table.

1,1	2,1		4,1	5,1	
1,2	2,2		4,2		6,2
1,3		3,3		5,3	
1,4	2,4			5,4	
1,5		3,5			
1,6	2,6		4,6		6,6

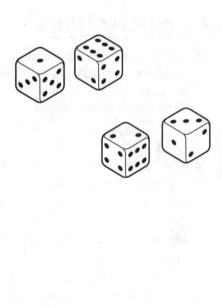

2a. How many of the outcomes in the table add up to an even number? _____

2b. For each toss, what is the probability of getting an even number? _____

3a. How many of the outcomes in the table add up to an odd number? _____

3b. For each toss, what is the probability of getting an odd number? _____

AFTER YOU PLAY:

4. How many tosses added up to an even number? _____

5. How many tosses added up to an odd number? _____

6. Play the game two more times. Record your results on a separate piece of paper.
   Do you think the game is fair? Explain why or why not.

_____

_____

Probability at Play

Students learn how probability is used in their favorite video games and practice writing probability as a fraction, decimal, and percent.

⟳→ Directions

1. Duplicate a copy of the reproducible for each student.

2. Review with students the process for converting a fraction to a decimal and a percent. Give a few examples of fractions and have volunteers give the equivalent decimals and percents. Use an example such as $\frac{5}{8} = \frac{62.5}{100} = 63\%$ so students can become familiar with rounding percents.

3. Discuss reasons why it might be convenient to look at probability as a decimal or a percent, rather than as a fraction. For example, it may be easier to compare probabilities when they are written as percents rather than as fractions with different denominators.

4. Distribute the reproducible and let students complete it on their own. You may want to let them use calculators.

☆ Taking It Farther

Ask students to look for other everyday areas in which probability is used, such as weather prediction. They may do research and then present their findings to the class.

✓ Assessing Skills

Observe whether students can successfully convert fractions to decimals and percents, and whether they use mental estimation to determine if their answers make sense. Some students, for example, may write $\frac{1}{2}$ as 2% or as 0.5%.

LEARNING OBJECTIVE

Students write probability as a fraction, decimal, and percent.

GROUPING

Individual

MATERIALS

✳ *Probability at Play* reproducible (p. 16)

✳ calculators (optional)

ANSWERS

START → $\frac{3}{4}$ → 75% → 0.4 → $\frac{2}{5}$ → 83% → 0.83 → 25% → $\frac{1}{4}$ → $\frac{41}{100}$ → 41% → $\frac{3}{8}$ → 0.375 → $\frac{14}{15}$ → 93% → 60% → $\frac{18}{30}$ → 0.26 → 26% → FINISH

Probability at Play

Flip on your favorite Sega or Nintendo game, and chances are you're also turning on a game of probability! To keep games interesting, video game designers program random chance into their games. That way, you'll never know exactly how many bad guys you'll have to battle, or just where that secret power source is hidden.

To help Luis get a high score in the video game below, use your math smarts to find the equivalent probability at each fork in the maze. Turn down the path of the correct probability to reach FINISH.

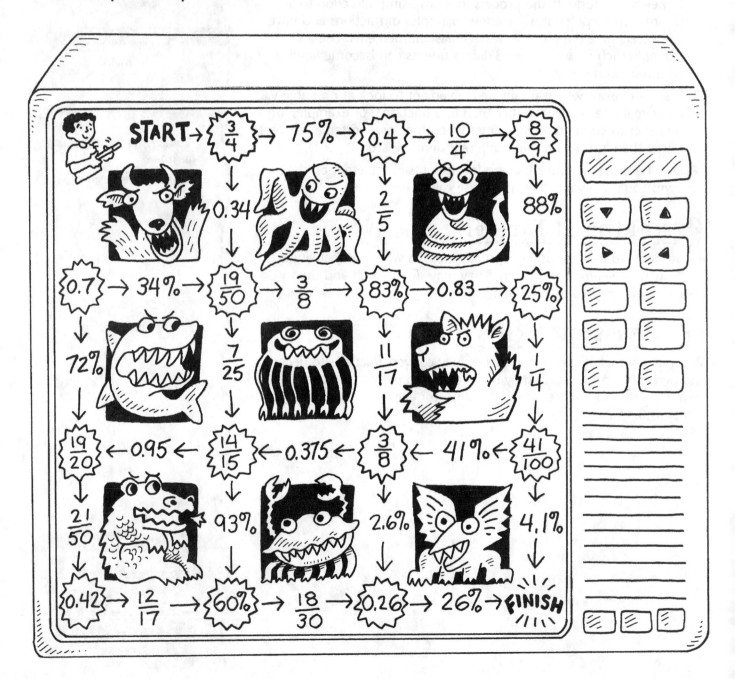

Go for a Spin!

Vrooom! Students practice finding the probability of a simple event as they motor around the track in a sports car board game.

⌖→ Directions

1. Review the formula for finding the probability of an outcome:

   Probability of event = P(event) = $\dfrac{\text{Number of Favorable Outcomes}}{\text{Number of Possible Outcomes}}$

2. If necessary, go over an example such as the following with the class:

   There are 5 slices of pizza in a box—2 pepperoni, 1 mushroom, and 2 eggplant. If you reach into the box without looking and take a slice, what is the probability you will get a pepperoni slice? [$\frac{2}{5}$]

3. Divide the class into groups of 3 or 4 and distribute one copy of the reproducible to each group.

4. Explain the rules of the game on the reproducible. Each student cuts out and colors a game marker from the reproducible. To play, students begin at the Start square. Each space tells which letter or pattern is needed to move ahead. At each turn, a player must decide which of the two spinners available provides the greater probability of landing on the desired outcome. First, the player writes each probability as a fraction. Then, the player compares the probability for Spinner A versus Spinner B, using scrap paper if necessary. Finally, the player spins, using the spinner that provides the greater probability. The first player to cross the finish line wins.

★ Taking It Farther

Have students use colored markers to draw their own spinners, filling in the sections with different patterns and colors. Divide students into pairs, and have each partner write the probability (in fraction form) of spinning a specific pattern or color on his or her partner's spinner. Students may then experiment by spinning the spinner 10 or 25 times, recording the outcomes, and writing a fraction to show how many times the desired outcome occurred. Students then compare the probability fraction and the outcome fraction.

✓ Assessing Skills

Note whether students are able to compare the values of two fractions with different denominators. You may review this skill by showing how to rewrite fractions with common denominators. Manipulatives and fractional diagrams can also help reinforce this skill.

LEARNING OBJECTIVE

Students find the probability of a simple event and write it as a fraction.

GROUPING

Cooperative groups

MATERIALS

For each group:
* *Go for a Spin!* reproducible (p. 18)
* scissors
* colored markers or pencils
* paper
* pencil and paper clip (to make the spinner)

Go for a Spin!

Rev up your probability skills as you race around the track below. Cut out the sports car game pieces and get ready to start your engines!

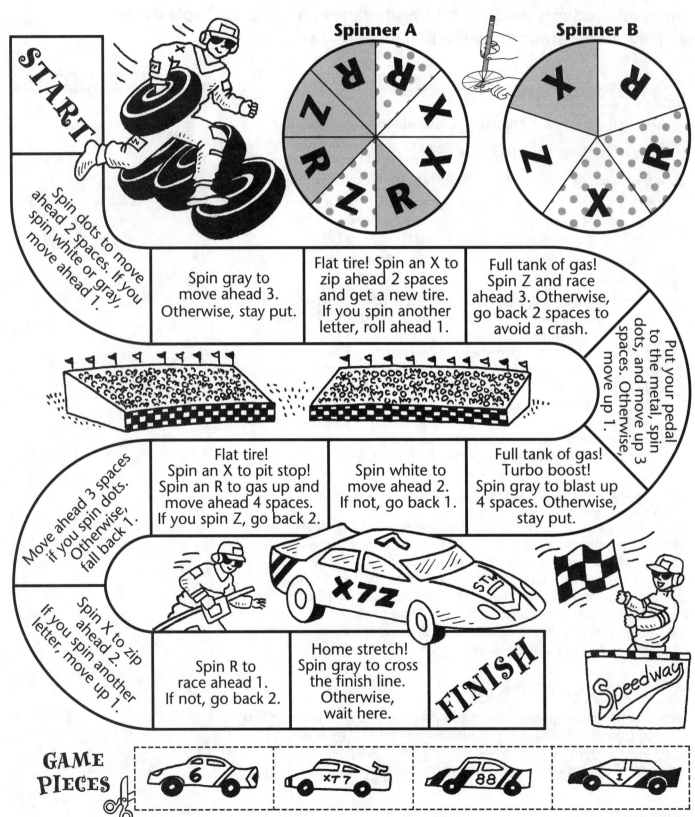

Spinner A

Spinner B

START

Spin dots to move ahead 2 spaces. If you spin white or gray, move ahead 1.

Spin gray to move ahead 3. Otherwise, stay put.

Flat tire! Spin an X to zip ahead 2 spaces and get a new tire. If you spin another letter, roll ahead 1.

Full tank of gas! Spin Z and race ahead 3. Otherwise, go back 2 spaces to avoid a crash.

Put your pedal to the metal, spin dots, and move up 3 spaces. Otherwise, move up 1.

Move ahead 3 spaces if you spin dots. Otherwise, fall back 1.

Flat tire! Spin an X to pit stop! Spin an R to gas up and move ahead 4 spaces. If you spin Z, go back 2.

Spin white to move ahead 2. If not, go back 1.

Full tank of gas! Turbo boost! Spin gray to blast up 4 spaces. Otherwise, stay put.

Spin X to zip ahead 2. If you spin another letter, move up 1.

Spin R to race ahead 1. If not, go back 2.

Home stretch! Spin gray to cross the finish line. Otherwise, wait here.

FINISH

Speedway

GAME PIECES ✂

Sweet Experiments

Probability—how sweet it is! Students use tasty candy manipulatives to compare theoretical and experimental probability.

➤ Directions

1. Distribute one paper bag to each pair or group of students, along with a small pile of Skittles or M&M candies.

2. Have students place several candies in the bag. There should be at least three colors represented, but they may use different numbers of candies for each color; for example, students may choose 5 green, 2 yellow, and 3 red candies.

3. Then they write down the probability of randomly picking each color from the bag, writing the probability as a fraction, decimal, and percent. In this case, the probability of picking green is $\frac{1}{2}$, or 0.5, or 50%.

4. One student holds the bag while another student reaches in without looking, picks out a candy, notes the color, and returns it to the bag. Students repeat this process over and over, keeping a tally of how many times each color is chosen. After 50 trials, students may use their tallies to figure out what fraction of the time each color was chosen, then write this experimental probability as a decimal and percent as well. (Students can round decimals to the nearest hundredth and percents to the nearest whole number.)

5. After comparing the theoretical and experimental probability, have students continue the process of picking candy until they have tallied 100 trials. They again compare the theoretical and experimental probabilities, and share results with the class.

✪ Taking It Farther

Repeat the experiment with odd numbers of candies (such as 9 or 11) so that students can practice working with more difficult fractions.

☑ Assessing Skills

Ask students: *When might it be an advantage to look at probability as a decimal or a percent?* [One possible answer: When you want to compare probabilities at a glance; for example, comparing a probability of $\frac{7}{9}$ and $\frac{24}{29}$ is more difficult than comparing the equivalent probabilities 78% and 83%.]

LEARNING OBJECTIVE

Students do hands-on experiments and write probability as a fraction, decimal, and percent.

GROUPING

Pairs or cooperative groups

MATERIALS

For each pair or group:
* large bag of Skittles or M&M's
* small paper bags
* paper and pencil

Take a Sample

Students take a sample of sample space.

⟳→ Directions

1. The *sample space* is the set of all possible outcomes for an experiment. In the case of flipping a coin, the sample space is heads, tails.Tell students that an *event* is always a subset of the sample space. For example, if they roll a number cube and want to roll a number greater than 3, the event is 4, 5, 6. The sample space is 1, 2, 3, 4, 5, 6. The ratio is:

$$P(\text{number} > 3) = \frac{3}{6} \quad \begin{array}{l}(4, 5, 6)\\ (1, 2, 3, 4, 5, 6)\end{array}$$

2. Take out your props and call on students to describe the sample space for tossing a number cube, picking a number from the bag, ordering a lunch or dinner entree at a restaurant, and sports teams playing against each other.

3. Then let students use the yellow pages to look up various listings such as "Locksmith." Ask: *What would be the sample space for calling a locksmith from the phone book?*

4. Finally, have students describe the sample space of all the math classes they might be assigned to next year (for example, Ms. Smith's class, Mr. Jones's class, and Mrs. Anderson's class).

★ Taking It Farther

Challenge students to choose a career such as astronaut or doctor and do research to find experiments that might occur while performing that job. They should do research to write a sample space for each experiment.

✓ Assessing Skills

Do students understand the difference among event, outcome, and sample space? To help visual learners, have them act out an experiment such as picking a number from the bag. Ask them to identify the event, outcome, and sample space.

LEARNING OBJECTIVE

Students are introduced to the concept of sample space.

GROUPING

Whole class

MATERIALS

* coin
* number cube
* paper bag containing strips of paper numbered from 1 to 10
* one or more of the following: takeout menus from local restaurants (lunch and dinner menus, if possible), the sports section of a newspaper, a telephone book yellow pages, a list of math teachers in your school for the grade following that of your class

Lost in Sample Space!

Students begin to see probability all around them as they find the sample space for occurrences in their everyday lives.

➤ Directions

1. Duplicate a copy of the reproducible for each student.

2. Review the terms *experiment*, *outcome,* and *sample space*. Ask students to name some experiments in their daily lives and some possible outcomes. Start them off with the example of dialing a phone number. They may hear a ring, a busy signal, or an answering machine message, or someone may answer the phone. With students' directions, make complete lists describing the sample space for each experiment.

3. If necessary, review a few examples of an experiment, an outcome, and a sample space, such as tossing a number cube. The experiment is tossing the cube; the outcome is that the cube turns up 3; the sample space is 1, 2, 3, 4, 5, 6.

4. Distribute the reproducible. You can call on volunteers to read the story at the top of the page out loud. Or have students silently read the story to themselves.

5. Let students complete the reproducible on their own. If they experience trouble, complete the first question as a class.

☆ Taking It Farther

Ask students to write their own short stories that involve various experiments. The stories may be about whatever students choose, or you may ask them to write stories about real or fictional trips. Classmates may read each other's stories and write out the sample space for each experiment in the stories they read.

✓ Assessing Skills

Watch for students who write an incomplete list of possible outcomes to describe the sample space.

LEARNING OBJECTIVE

Students practice applying the concept of sample space.

GROUPING

Individual

MATERIALS

* *Lost in Sample Space!* reproducible (p. 22)
* paper and pencils
* number cube (optional)

ANSWERS

1a. 2:00 A.M.

1b. 1:00, 2:00, 3:00, 4:00, 5:00, 6:00, 7:00, 8:00, 9:00, 10:00, 11:00, or 12:00 A.M.; 1:00, 2:00, 3:00, 4:00, 5:00, 6:00, 7:00, 8:00, 9:00, 10:00, 11:00, or 12:00 P.M.

2a. pork chops

2b. macaroni and cheese, pork chops, cotton candy

3a. The spaceship lands.

3b. Mars

4a. The robot brings back rocks from Mars.

4b. 5, 10, 15, 20, 25, 30, 35, 40, 45, 50, 55, 60, 65, 70, 75, 80, 85, 90, 95, 100

Lost in Sample Space!

Astronaut Andrea's space ship is malfunctioning! She's adrift somewhere in our solar system. Before she makes it back to Earth, she must deal with some weird random events on her ship.

Read about one of Andrea's days in space. Then answer the questions about sample space on a separate sheet of paper.

BRRRRRIIING! Andrea awoke with a start as her space alarm went off at 2:00 A.M., Eastern Standard Time. Oh, well, she thought, slapping down the snooze button. At least it was morning. She never knew what time the alarm would go off—just that it would be exactly on the hour, like 7:00 or 8:00. Yesterday, the alarm didn't go off until 5:00 P.M., and she missed the whole day!

Andrea floated over to the food console and pushed a button marked "breakfast." She wasn't too surprised when pork chops appeared. Ever since the ship started malfunctioning, breakfast had been a random selection of macaroni and cheese, pork chops, or cotton candy. Andrea sighed and wished for a banana muffin.

Cheering up, Andrea prepared for the best part of her day. Every afternoon, the ship would choose one of the 9 planets in the solar system and come in for a landing. Would today be Earth? The ship drew closer and closer . . . and shot off toward Mars. Arggh!

When she reached Mars, Andrea decided to pick up some rock samples for scientific research. She sent the ship's robot out to pick up 5 rocks. Unfortunately, the robot decided to bring any number of rocks that was a *multiple* of 5. Later, as Andrea tried to squeeze past the 85 rocks in her living quarters, she thanked her lucky stars that the machine couldn't carry more than 100 rocks. Falling asleep on a large boulder, Andrea decided to give up on her research until she got a better spaceship.

1. Experiment: The ship's alarm clock randomly picks a time to ring.
   a. What was the outcome for this experiment in the story?
   b. What is the sample space for this experiment?

2. Experiment: Andrea orders breakfast.
   a. What was the outcome for this experiment in the story?
   b. What is the sample space for this experiment?

3. Sample space: Mercury, Venus, Earth, Mars, Jupiter, Saturn, Uranus, Neptune, Pluto
   a. What was the experiment for this sample space in the story?
   b. What was the outcome in the story?

4. Outcome: 85
   a. What was the experiment for this outcome in the story?
   b. What is the sample space for this outcome?

Clothes Combos

How many outfits can students make from a pile of clothes? With tree diagrams, they'll always dress for success!

Directions

1. Duplicate a copy of the reproducible for each student.

2. At the start of class, bring out your items of clothing. Make sure the different types of items are all mixed together. Ask volunteers to come up and make as many different outfits as possible using one hat, one shirt, and one pair of pants for each outfit. Choose another volunteer to record all the different outfits on the board.

3. When students are confident they have come up with every possibility, distribute the reproducible. Let students complete it on their own.

4. After students have finished, have them compare answers.

5. Ask volunteers to draw tree diagrams on the board showing how many outfits can be made with the items of clothing they looked at earlier. Students may draw each item or describe it in words. When the volunteers are finished, determine whether they have come up with the same number of outfits as before.

6. Ask students to describe the benefits of using a tree diagram versus making a list. In this case, a tree diagram makes it easier to keep track of every possible outcome, and every outcome is clearly shown in order.

Taking It Farther

Repeat the activity several times using new groups of clothes. You may also ask students to make a list of several of their favorite clothing items and draw tree diagrams to count how many outfits they can make.

Assessing Skills

Observe whether students are able to correctly describe all the outcomes from their tree diagrams. If they are having difficulty, try having them trace the "limbs" of a tree diagram with their finger. For example, they may start at "green hat," and follow the branch to "yellow shirt" and then "blue pants."

LEARNING OBJECTIVE

Students draw tree diagrams to count possible outcomes.

GROUPING

Whole class/Individual

MATERIALS

* *Clothes Combos* reproducible (p. 24)

* a number of different items of clothing: hats, shirts, and pants, for example (These may be funny/silly items if you want.)

ANSWERS

A. 6

B. 2

C. 3

Check students' completed tree diagrams. Outcomes: PLSN, PLSa PShSn, PShSa, StLSn, StLSa, StShSn, StShSa

D. 4

E. 4

F. 4

Name _____ Date _____

Clothes Combos

How many combinations can you make with your favorite clothes? A tree diagram can show you. For example, Doug has one baseball cap, three shirts, and two pairs of pants. If he chooses one hat, one shirt and one pair of pants for each outfit, how many outfits can he make?

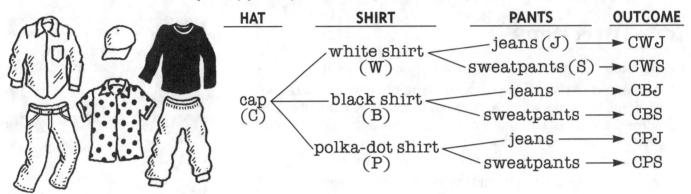

HAT	SHIRT	PANTS	OUTCOME
	white shirt (W)	jeans (J)	CWJ
		sweatpants (S)	CWS
cap (C)	black shirt (B)	jeans	CBJ
		sweatpants	CBS
	polka-dot shirt (P)	jeans	CPJ
		sweatpants	CPS

A. How many of Doug's outfits include a baseball cap? _____

B. How many outfits include a white shirt? _____

C. How many outfits include jeans? _____

Fill in this tree diagram to find out which different outfits Stella can make with *her* clothes. She can pick one shirt, one skirt, and one pair of shoes for each outfit. Here's what she's got: polka-dot shirt, striped shirt, long skirt, short skirt, sneakers, and sandals.

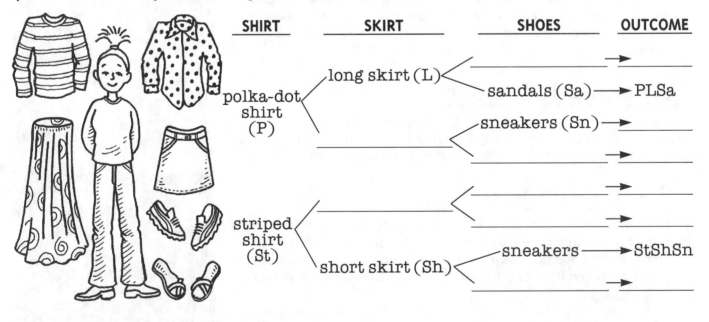

SHIRT	SKIRT	SHOES	OUTCOME
	long skirt (L)		→
polka-dot shirt (P)		sandals (Sa)	PLSa
		sneakers (Sn)	→
	_____		→
			→
striped shirt (St)	_____		→
	short skirt (Sh)	sneakers	StShSn
			→

D. How many of Stella's outfits include a striped shirt? _____

E. How many outfits include a long skirt? _____

F. How many outfits include sneakers? _____

Tree-Licious Diagrams

Students use tree diagrams to find funky food combinations in your school cafeteria, a deli, or even an ice cream store.

⟳→ Directions

1. Review tree diagrams with the class.

2. Using the menus, pick out different food items and have students figure out how many combinations could be created with the options listed. Here are some possible examples:

   ✳ At a pizzeria: If you have 5 toppings to choose from and 3 pizza sizes to choose from, how many different pizzas could you make if each pizza has one topping?

   ✳ At a deli: How many different sandwiches could be made using 1 kind of bread, 3 meats, and 2 cheeses?

   ✳ At the school cafeteria: Find all the possible combinations that contain 1 drink, 1 sandwich or entree, and 1 dessert.

   ✳ At a restaurant: Pick a few appetizers, entrees, beverages, and desserts. How many combinations can be made with these choices?

☆ Taking It Farther

To prepare students to learn the Fundamental Counting Principle, challenge them with the following problem: *You work at an ice cream store that offers 31 flavors of ice cream, 25 kinds of flavored syrup, and 18 different toppings. A customer asks how many different types of sundaes could be made using one kind of ice cream, one syrup, and one topping. Could this problem be solved with a tree diagram?* [yes] *Would this approach be practical?* [Probably not, because the problem would take a long time and would require a *very large* piece of paper.]

✔ Assessing Skills

Note whether students completely draw every section of a tree diagram, especially if there are two choices in the first column.

LEARNING OBJECTIVE

Students draw tree diagrams to count possible combinations of food.

GROUPING

Whole class

MATERIALS

✳ menus from the school cafeteria, a local deli, ice cream store, pizzeria, or any other food store (You may invent food choices for this activity if menus are not available.)

Count on Sports Schedules

Probability theory makes a touchdown in students' minds as they learn how the Fundamental Counting Principle can help simplify a school's sports schedule.

➤ Directions

1. Start with an example from the WNBA (Women's National Basketball Association). Tell students that in the 1997 season, there were 8 teams in the WNBA. Ask students: *If each team must play one away game at each of the other 7 teams' stadiums, how many games must be played?*

2. Explain that they could use tree diagrams to find out this information, but it would take a long time since there are so many outcomes. Tell students that an easier way is to use the Fundamental Counting Principle. Write the following definition on the board and discuss it with the class:

   Fundamental Counting Principle
   If one choice can be made in x ways, and for each of these choices a following choice can be made in y ways, then the choices can be made in xy ways (that is, x times y ways).

3. In the WNBA example, there are 8 teams, and each team must play at 7 other teams' stadiums. Therefore, there will be 8×7, or 56, games played in all.

4. Students may check the answer by drawing tree diagrams for each team's away games and adding the results. Then use your own school's sports teams for additional examples. For instance, if there are 6 schools in your soccer league, and each soccer team must play an away game at every other school, how many games will be played?

☆ Taking It Farther

Challenge students to use the Fundamental Counting Principle to find the following: the total number of possible area codes; the total number of possible 7-digit telephone numbers; the total number of possible car license plates, where each plate has two letters followed by three numbers. (Students may use calculators to do the arithmetic.)

✔ Assessing Skills

Ask students to use the Fundamental Counting Principle to find the number of possible sundaes in Taking It Farther on page 25. [$31 \times 25 \times 18 = 13,950$ sundaes]

LEARNING OBJECTIVE

Students use the Fundamental Counting Principle to count outcomes.

GROUPING

Whole class

MATERIALS

* sports schedules for different varsity sports in your school or a sports schedule from a professional league
* calculators (optional)

Pet Peeves

Some wild and woolly manipulatives will help students model and understand the Fundamental Counting Principle.

⟳→ Directions

1. Duplicate and distribute the reproducible to students. Direct them to cut along the dotted lines to separate each animal card.

2. Pose the following problem to students: *Wally loves animals. Today he brought home 3 dogs, 2 cats, and 1 cow! But Wally's mom says he can keep only 1 of each type of animal for a pet. How many different ways can Wally choose 1 dog, 1 cat, and 1 cow?*

3. To answer the question, ask students to pick out 3 dog cards, 2 cat cards, and 1 cow card. Have them move around the cards to model the answer to the question. Write down the combinations they discover. Then ask them to sort the cards into one vertical dog column, one cat column, and one cow column.

4. Remind students that to use the Fundamental Counting Principle, they multiply the number of items in each column: $3 \times 2 \times 1 = 6$ possible combinations. This answer should be the same as the number of combinations you wrote on the board.

5. Next, have students turn all their cards facedown. After they mix the cards, ask students to pick out 8 different cards. On separate sheets of paper, students write down which cards they picked and use the Fundamental Counting Principle to figure out how many different combinations could be made that contain one animal from each type picked. For example, of the 8 cards picked, there could be 2 dogs, 3 cats, 1 fish, 2 gerbils. To determine the number of combinations that could be made, multiply $2 \times 3 \times 1 \times 2$ for 12 combinations. Repeat this exercise several times, asking students to pick a different number of cards each time.

★ Taking It Farther

Let students go back and use the Fundamental Counting Principle to check their work for the tree diagram activities on pages 23–25. The outcomes should be the same. Ask students: *Which method of counting outcomes do you prefer? Why?*

✓ Assessing Skills

Ask students: *Say you pick 8 animal cards. If you pick 3 gerbil cards and 5 fish cards, will you have the same number of possible combinations as if you pick 1 gerbil card and 7 fish cards? Why or why not?*

LEARNING OBJECTIVE

Students use manipulatives to model the Fundamental Counting Principle.

GROUPING

Individual

MATERIALS

* *Pet Peeves* reproducible (p. 28)
* paper and pencil
* scissors

Name _____   Date _____

Pet Peeves

Cut out the animal cards.
Group the same types of animals together.

Bowser	Katrina	Jaws	Lance	Gertrude
Dotty	Santa Claws	Bubbles	Harry	Moo-ria
Wags	Snaggles	Rover	Jane	Daisy
Lulu	Tiger	Glub-Glub	Nibbles	Henrietta
Rex	Snowball	Fluffy	Farley	Babs

Probability Scholastic Professional Books

School Daze

Students choose whether to use tree diagrams or the Fundamental Counting Principle to help a girl make choices during her school day.

Directions

1. Duplicate a copy of the reproducible for each student.

2. Distribute the reproducible. Review tree diagrams and the Fundamental Counting Principle as necessary.

3. Explain to students that they may use either tree diagrams or the Fundamental Counting Principle to solve each problem on the reproducible. There is no right or wrong method to choose, but students should be aware that tree diagrams may take up a lot of space and time where many choices are involved.

4. Allow students to complete the reproducible on their own. Remind them to answer the questions and show their solutions on separate sheets of paper.

5. Have students share their answers with the class and explain why they chose each method to solve the problems.

Taking It Farther

Encourage students to write their own problems based on choices they make during their school day. The problems can be based on real choices, or students may invent situations.

Assessing Skills

Observe students who choose to use just one method (tree diagrams or the Fundamental Counting Principle) to solve the problems. To determine whether they understand how to use the other method, suggest that they check a few of the answers using the other method.

LEARNING OBJECTIVE

Students compare the merits of using tree diagrams and the Fundamental Counting Principle to count outcomes.

GROUPING

Individual

MATERIALS

* *School Daze* reproducible (p. 30)
* paper and pencil

ANSWERS

1a. 9 ways

1b. Answers may vary.

1c. Sample answer:

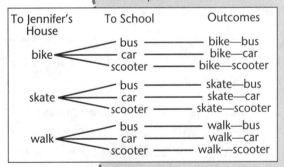

2a. 343 combinations

2b. Answers may vary.

2c. Sample answer: 7 x 7 x 7 = 343 combinations

3a. 1,920 combinations

3b. Answers may vary.

3c. Sample answer: 12 x 4 x 8 x 5 = 1,920 combinations

School Daze

Wendy always likes to make an informed decision. But how many choices does she have as she goes through her school day?

Help Wendy decide by drawing a tree diagram or using the Fundamental Counting Principle. Choose whichever method you like!

1. Wendy plans to travel to school with her friend Jennifer. First, Wendy needs to get to Jennifer's house. To get there, she can ride her bike, in-line skate, or walk. From there, the girls can get to school by taking the bus, getting a ride in Jennifer's dad's car, or riding on Jennifer's scooter.

   a. How many different ways can Wendy get to school in all?

   _____

   b. Which method did you use to solve this problem? Why?

   _____

   _____

   _____

   c. Show your solution on a separate sheet of paper.

2. Lunchtime! Wendy eats a sandwich in the cafeteria and decides to go to the ice cream parlor across the street. Wendy has 7 favorite ice cream flavors: chocolate, strawberry, cookies 'n' cream, vanilla, mint, peanut butter swirl, and peach.

   a. If Wendy gets a triple-scoop cone, how many combinations can she make using her favorite flavors?

   _____

   b. Which method did you use to solve this problem? Why?

   _____

   _____

   _____

   c. Show your solution on a separate sheet of paper.

3. For English homework, Wendy has 4 lists of poems. She must pick a poem from each list and read it. The first list contains 12 poems. The second list has 4 poems, the third list has 8 poems, and the fourth list has 5 poems.

   a. How many different combinations of poems could Wendy choose to read?

   _____

   b. Which method did you use to solve this problem? Why?

   _____

   _____

   _____

   c. Show your solution on a separate sheet of paper.

Gold, Silver, Bronze

How many different ways can athletes place in races or other competitions? Students find out by working with permutations!

◎→ Directions

1. Go over the concept of permutation with the class. A *permutation* is an arrangement or list of choices in a particular order. (In combinations, which can be found with tree diagrams or the Fundamental Counting Principle, order is not important.)

2. Pose this problem to the class: *There are 4 runners in a race. How many different ways can the runners place first, second, and third?*

3. Explain that to answer this question, you would write the following permutation problem: $P(4, 3)$. This means that you are finding the permutation of 4 things (in this case, runners) taken 3 at a time (in this case, first, second, and third place).

   To find the answer, you multiply as follows: $4 \times 3 \times 2$. Any of 4 runners may get first place; any of the remaining 3 runners may get second place (because 1 runner has already won first place), and any of the remaining 2 runners may get third place (because 2 runners have already won first and second place).

   Therefore, there are 24 ways for the runners to place first, second, and third. Point out to students that the number of choices decreases by 1 each time.

4. Now have students use the lists of athletes to write and solve more permutation problems. For example, if there are 10 figure skaters in the Olympics, how many ways can they win the gold, silver, and bronze medals? [$P(10, 3)$; $10 \times 9 \times 8 = 720$]

☆ Taking It Farther

Challenge students to use the lists of athletes to find the total number of ways athletes could place in a competition. For example, to find out how many different ways 10 skaters could place in the Olympics, you would solve the problem $P(10, 10)$. The symbol 10! can also be used to express this problem. The notation 10! is read "ten factorial." It means $10 \cdot 9 \cdot 8 \cdot 7 \cdot 6 \cdot 5 \cdot 4 \cdot 3 \cdot 2 \cdot 1$.

✔ Assessing Skills

To see whether students understand permutations, have them use manipulatives (such as scraps of paper with athletes' names) to model problems. For instance, ask: *How many ways can 3 runners place first and second in a race?* [$P(3, 2) = 3 \times 2 = 6$]

LEARNING OBJECTIVE

Students are introduced to permutations and use this concept to solve problems.

GROUPING

Whole class

MATERIALS

✳ scrap paper and pencil

ADVANCE PREPARATION

Compile lists of athletes who compete in popular sporting events such as the Olympics or tennis or golf tournaments. Otherwise, fictional athletes may be used.

Career Day Conundrums

Students get up close and personal with permutations as they figure out how many career presentations can be seen at Mathville Middle School's Career Day.

Directions

1. Duplicate a copy of the reproducible for each student.

2. Review permutations with the class if necessary. Here is an example you can use: *A radio disc jockey (DJ) wants to pick 3 songs from a list of 5 songs to play on the air. He doesn't want to repeat any songs. How many ways can he play 3 songs from the list?*

3. The problem would be written as follows: $P(5, 3)$. In other words, you are finding the permutation of 5 things taken 3 at a time. For the first song, the DJ has 5 choices. For the second song, he has 4 choices, since he has played 1 song from the list already. For the third song, he has 3 choices, since he has played 2 songs from the list already. So to find the answer, you would multiply $5 \cdot 4 \cdot 3$. There are 60 ways the DJ could play 3 songs from the list.

4. Let students complete the problems on their own. When students are finished, you may want to have them work in pairs to check and discuss their answers.

Taking It Farther

Have students use their own class schedule to find the number of permutations of ways they could take their classes. For instance, if there are 8 class periods in a day, and they take 7 classes, how many different ways could they arrange their schedules? Tell students to assume that they can take any class at any time of day.

Assessing Skills

Note whether students understand how many numbers to multiply to find the correct permutation. For example, in trying to find the number of ways to play 3 songs from a 5-song list, some students may multiply $5 \cdot 4 \cdot 3 \cdot 2 \cdot 1$. This would be incorrect, since the DJ wants to play only 3 songs at a time.

LEARNING OBJECTIVE

Students practice using permutations to find arrangements where order is important.

GROUPING

Individual

MATERIALS

* *Career Day Conundrums* reproducible (p. 33)
* pencil and scrap paper

ANSWERS

1a. $P(5, 4)$

1b. $5 \times 4 \times 3 \times 2 = 120$ ways

2a. $P(10, 2)$

2b. $10 \times 9 = 90$ ways

3a. $4 \times 3 \times 2 \times 1 = 24$ ways

3b. $4 \times 3 \times 2 = 24$ ways

4a. Answers may vary.

4b. Answers may vary depending on number of presentations chosen, but permutation problem should be set up as: $P(n, 4)$, where n = number of presentations.

4c. Answers may vary, but permutation problem should be set up as: $P(n, 2)$.

Career Day Conundrums

It's Career Day at Mathville Middle School! Kids at the school will participate in workshops about many exciting jobs. Unfortunately, they can't attend every workshop. Help the students find out how many ways they can choose to attend different career workshops.

Use the tables and your knowledge of permutations to answer the questions. Remember, none of the students wants to attend the same workshop more than once.

Table One: CAREER DAY WORSHOPS	
Firefighter	Interior Decorator
Engineer	Politician
Doctor	Lifeguard
Hotel Manager	Telephone Repairer
Journalist	Photographer

Table Two: WORKSHOP TIMES
All presentations are offered at each time listed.
9:30 A.M.
10:30 A.M.
1:30 P.M.
2:30 P.M.

1. Victor is interested in the following workshops: Politician, Journalist, Engineer, Doctor, and Hotel Manager. If he attends one workshop at each of the four times they are offered, how many different ways can he arrange his schedule?

   a. Write this as a permutation problem.

   _____

   b. How many ways can Victor arrange his schedule?

   _____

2. Lydia is interested in all of the workshops, but she is available to attend workshops only in the morning. How many ways can she arrange her schedule?

   a. Write this as a permutation problem.

   _____

   b. How many ways can Lydia arrange her schedule?

   _____

3. Alex has chosen 4 workshops: Lifeguard, Firefighter, Photographer, and Engineer.

   a. If Alex is available for all 4 workshop times, how many ways can he arrange his schedule?

   _____

   b. If he is available at only 3 of the times, how many ways can Alex arrange his schedule?

   _____

4. a. Which of the workshops interest you?

   _____

   b. If you could attend all 4 workshop times, how many ways could you arrange your schedule?

   _____

   c. Suppose you could attend only afternoon workshops. How many ways could you arrange your schedule?

   _____

Turnover Turns

Fractions and an area model will help students understand the twists and turns in these pastry probability problems.

⟶ Directions

1. Duplicate and distribute the reproducible. To familiarize students with how to use an area model, go over the first problem with the class. You may want to draw the maze and the area model on the board so that you can refer to it during the class discussion.

2. Start by tracing each path from the left. Ask students:
   * *At the first fork in the path, what is the probability that Terence will choose each path?* [$\frac{1}{3}$]
   * *When the upper path divides into two more paths, what is the probability for each of these new paths?* [half the probability of the previous path, or $\frac{1}{6}$ each]

3. Ask a volunteer to color in the fractions in the area model with colored chalk. A different color should be used for each destination to show even more clearly how the fractions represent the destinations.

4. Have students add the fractions to find the probability that Terence will end up at each destination on the maze. Write in the answers to questions 1 and 2.

5. Divide the class into cooperative groups and have them draw area models to answer the remaining questions, using crayons or colored pencils to color in their models.

☆ Taking It Farther

When the groups have completed the reproducibles, challenge them to draw their own mazes on separate paper. Then, groups switch mazes and draw area models to find the probabilities of randomly choosing each path and of ending up at each finishing location. Students may invent funny or interesting scenarios to go along with their problems.

✓ Assessing Skills

Some students may try to calculate the probability of reaching each destination by counting the number of path endings that reach that destination. Point out that this approach will not work, since the probability of reaching a path ending depends on how many times the path has forked along the way.

LEARNING OBJECTIVE

Students use fractions and area models to understand the probability of compound events.

GROUPING

Cooperative groups

MATERIALS

* *Turnover Turns* reproducible (p. 35)
* colored chalk
* crayons or colored pencils

ANSWERS

1. $\frac{1}{6} + \frac{1}{6} = \frac{2}{6}$, or $\frac{1}{3}$

2. $\frac{1}{6} + \frac{1}{6} + \frac{1}{3} = \frac{4}{6}$, or $\frac{2}{3}$

3. $\frac{1}{6} + \frac{1}{4} = \frac{5}{12}$

4. $\frac{1}{6} + \frac{1}{6} + \frac{1}{4} = \frac{7}{12}$; area model for Maze 2: $\frac{1}{6} + \frac{1}{4} = \frac{5}{12}$.

Upper Path	Wild Woods	$\frac{1}{6}$
	Wild Woods	$\frac{1}{6}$
	Ricardo's Cabin	$\frac{1}{6}$
Lower Path	Ricardo's Cabin	$\frac{1}{4}$
	Wild Woods	$\frac{1}{4}$

Turnover Turns

Every Tuesday, Terence delivers his tasty turnovers. Today's deliveries will take him through some long, winding forest roads to Carla's Campground. Unfortunately, Terence lost the directions to the campground! If he just guesses which way to turn at each fork in the road, what is the probability he will wind up in the right place?

Use the area model to help you find out!

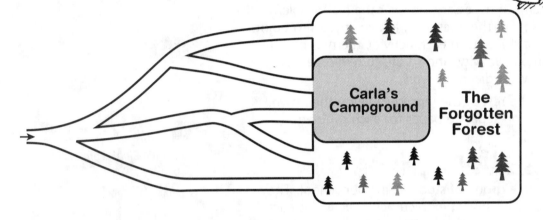

Upper Path	Forest
	Carla's
Middle Path	Forest
	Carla's
Lower Path	Forest

Area Model of MAZE 1

MAZE 1: Terence needs to get to Carla's Campground.

1. What is the probability Terence will end up at Carla's Campground if he randomly chooses which way to turn at each path? _____

2. What is the probability he will wind up in the Forgotten Forest? _____

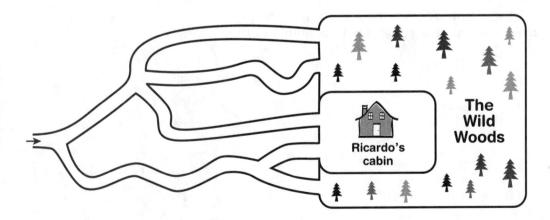

MAZE 2: Terence needs to get to Ricardo's cabin.

3. If he randomly chooses which paths to turn down, what is the probability Terence will make it to the cabin? _____

4. What is the probability he will end up in the Wild Woods? _____

Join the Club!

When looking at a group of kids and their extracurricular activities, Venn diagrams make visualizing probability extra clear.

➤ Directions

1. Distribute a copy of the reproducible to each student. Start by reviewing Venn diagrams. If necessary, go over the first problem with the class. Before you answer the questions, ask volunteers to point out the total number of students in each activity, the number of students in more than one activity, and so on, to reinforce an understanding of how Venn diagrams work.

2. Let students complete the rest of the reproducible on their own, drawing Venn diagrams on separate paper for each problem.

☆ Taking It Farther

Have your class survey a large group of students in your school. They can ask several questions, such as what extracurricular activities each person participates in, favorite cafeteria lunches, and so on. Students can then draw Venn diagrams based on their research and use the diagrams to make probability statements about the student body; for example, "If I picked a student at random, the probability that pizza is that person's favorite lunch would be 20 percent."

☑ Assessing Skills

Note whether students check their Venn diagrams by adding up all the numbers in each circle to make sure the sum equals the total number of students in that club.

LEARNING OBJECTIVE
Students draw Venn diagrams to find probabilities.

GROUPING
Individual

MATERIALS
* *Join the Club!* reproducible (p. 37)
* pencil and paper

ANSWERS

1a. 35

1b. 20

1c. 35

1d. $\frac{50}{100}$, or $\frac{1}{2}$

1e. $\frac{20}{100}$, or $\frac{1}{5}$

1f. $\frac{35}{100}$, or $\frac{7}{20}$

2a. 75

2b. $\frac{60}{100}$, or $\frac{3}{5}$

2c. $\frac{25}{100}$, or $\frac{1}{4}$

2d. $\frac{65}{100}$, or $\frac{13}{20}$

2e. $\frac{0}{100}$, or 0

3a. $\frac{25}{50}$, or $\frac{1}{2}$

3b. $\frac{10}{50}$, or $\frac{1}{5}$

10

Dandelion 15 | 15 | Plaid 10

Join the Club!

The kids at A.K. Tivitee Junior High love after-school clubs. Which clubs are
they most likely to join? Check out some Venn diagrams to get a closer look.

1. In one group of 100 students, there are lots of members
   of the Harmonica Club and the Tiddleywinks Society.
   Some students even belong to both clubs!

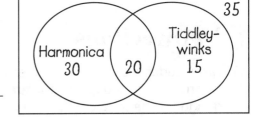

   a. According to the Venn diagram, how many
      students are in the Tiddleywinks Society? _____

   b. How many students are in both clubs? _____

   c. How many students are not members of either club? _____

   d. If you pick one of the 100 students randomly, what is the
      probability that he or she is a member of the Harmonica Club? _____

   e. What is the probability that the student is a member of both clubs? _____

   f. What is the probability that the student isn't be in either club? _____

2. It was discovered that many of the same students also play on the varsity
   Foosball team. Many students enjoy Cheese Sculpture Classes as well.

   a. How many students play foosball? _____

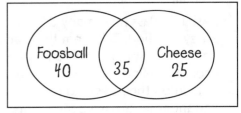

   b. If you pick a student at random,
      what is the probability that he or
      she takes cheese sculpture classes? _____

   c. What is the probability that you will pick a student who
      takes cheese sculpture classes but does not play foosball? _____

   d. What is the probability that you will pick a student who plays foosball
      or takes cheese sculpture classes but does not do both activities? _____

   e. What is the probability you will pick a student who
      doesn't participate in either foosball or cheese sculpture? _____

3. In a different group of 50 students, there are 30 members of the
   Dandelion Conservation Organization, and 25 members of Students for
   the Promotion of Plaid. Of those students, 15 belong to both groups.
   On a piece of paper, draw a Venn diagram to illustrate this group of students.

   a. If you pick a student at random, what is the probability
      he or she belongs to Students for the Promotion of Plaid? _____

   b. What is the probability he or she doesn't belong to either club? _____

Stamp to It!

How many different ways could you tear off four attached stamps from a sheet? Don't look to your mail carrier for the answer—make a model!

➔ Directions

1. Have students cut out four paper squares of identical size so they can be arranged in groups of four later. They should create a stamp design for the squares. The design can be simple, as long the top and bottom of it are clear. Students should draw their designs on all four squares.

2. Draw the following shape on the board:

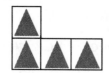

   Explain to students that the drawing shows one arrangement in which four stamps could be torn from a sheet of stamps so that they remain attached.

3. Ask the class how many different arrangements for the four stamps they can find. Tell them that each stamp must be attached to at least one other stamp along an entire side—diagonal attachments don't count. To find the answer, students move around the stamps they made on their desks to make different arrangements. When new arrangements have been found, students sketch them on separate sheets of paper.

⭐ Taking It Farther

Add a fifth stamp to the activity. Challenge students to see who can come up with the most arrangements.

✓ Assessing Skills

Note whether students realize that most of the shapes can be arranged in more than one way by rotating the shape 90°.

LEARNING OBJECTIVE

Students work with manipulatives to find arrangements.

GROUPING

Individual

MATERIALS

✳ drawing paper
✳ pencils
✳ crayons or colored markers
✳ scissors

ANSWERS

Spin to Win

It's the bottom of the ninth—and a simulation experiment can show students how likely it is that they'll win the game.

⟶ Directions

1. Divide the class into groups and distribute a reproducible to each group. Briefly review fractions and circle graphs with students. Show how fractional probabilities are translated into slices of the circle graphs to make the spinners in the activity.

2. Let each group conduct the simulations independently, spinning a paper clip around a pencil point on each spinner. Before spinning, the group predicts whether each player will score for the win. Each member of the group can take turns spinning.

3. When all the groups are finished, have the class share and discuss their answers.

★ Taking It Farther

✳ Let students create their own sports scenarios and draw circle graphs to illustrate the scenarios. They may even research their favorite sports players (to find out a basketball player's free throw shooting percentage, for example) and use that information to make their own circle graphs to use in simulations.

✳ For the basketball problem where the player must make two shots, ask students to multiply the probabilities to find the mathematical probability that the player will score on both shots. Then have them compare that probability to the result of their simulations. Tell students to repeat the simulation several more times and note whether the predicted probability is now more accurate.

✓ Assessing Skills

Ask students to look at the first spinner to answer these questions:

✳ *For Lefty's next 12 times at bat, how many times can she expect to make an out?* [3]

✳ *How do you know?* [She makes an out $\frac{1}{4}$ of the time, and 3 is $\frac{1}{4}$ of 12.]

LEARNING OBJECTIVE

Students conduct simulations to gain a concrete understanding of fractional probability.

GROUPING

Cooperative groups

MATERIALS

For each group:

✳ *Spin to Win* reproducible (p. 40)

✳ pencil and paper clip (to make the spinner)

Spin to Win

These athletes are in a tough spot. At the end of a game, it's up to them to make the winning score!

Based on his or her previous scoring record, we know the probability that each athlete will score. What does that mean for the team? Make a prediction based on the circle graph. Then, for each problem, spin the spinner to see what happens. Repeat the simulation 10 times for each problem. Mark your results in the Score Boxes.

1 It's the bottom of the ninth inning, and Lefty is up at bat. For her team to win the game, Lefty must hit a home run or a triple to drive in other runners on the bases. What are her chances? Over the past season, she's hit a home run $\frac{1}{16}$ of her times at bat. She's hit a triple $\frac{1}{8}$ of the time, a double $\frac{3}{16}$ of the time, a single $\frac{3}{8}$ of the time, and has made an out $\frac{1}{4}$ of the time. To see what Lefty does this time, spin the spinner. Record your result in the Score Box.

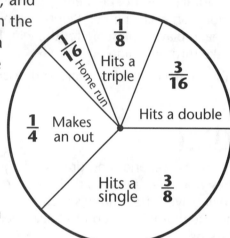

SCORE BOX

Trial	Out, Single, or Double	Triple or Home Run
1		
2		
3		
4		
5		
6		
7		
8		
9		
10		

2 Rhonda's basketball team is counting on her. It's the end of the game, and Rhonda is at the freethrow line. She has two shots, and she must make both of them to win the game. Will Rhonda do it? In the past, she has made $\frac{2}{3}$ of her free throw shots and missed $\frac{1}{3}$. Spin the spinner once to find out if Rhonda makes her first shot, then spin again to see if she makes the second shot. Repeat the simulation 10 times and make a check for each basket she makes in the Score Box. If she misses, leave the box blank.

SCORE BOX

Trial	First Shot	Second Shot
1		
2		
3		
4		
5		
6		
7		
8		
9		
10		

False Hopes

Should students study for that true or false quiz, or just try to guess the answers? After completing this simulation, they'll run to hit the books!

⟳→ Directions

1. On the board, write the following, in vertical columns:

1. T	6. F	11. F	16. T	21. T
2. T	7. F	12. T	17. F	22. F
3. T	8. F	13. F	18. F	23. T
4. T	9. F	14. T	19. T	24. F
5. T	10. F	15. F	20. F	25. T

2. Ask students to copy the list of numbers and T or F answers in one column down the left-hand side of a piece of paper. While they complete this task, hand out the coins.

3. Now have students flip their coins to see if they would guess correctly on each of the questions. Explain that flipping heads will represent True, while flipping tails will represent False. As they flip their coins, they mark a check next to each answer that the coin flip matches and an X next to each answer that the coin flip does not match.

4. When they are finished, students may score 4 points for each answer they matched with a coin flip. Have students determine their "grade" and compare their results.

☆ Taking It Farther

Repeat the simulation, this time using multiple-choice questions with A, B, and C as possible answers. Write a list of 10 or 20 numbers with A, B, and C answers on the board. Have each student conduct the simulation by rolling a number cube; 1 and 2 can represent A, 3 and 4 can represent B, and 5 and 6 can represent C. What difference do students see in their results this time?

✔ Assessing Skills

Ask students: *Does the chance of getting 100 percent by guessing on the quiz increase or decrease as more questions are added? Why?*

LEARNING OBJECTIVE

Students use a simulation experiment to get a hands-on understanding of probability of compound events.

GROUPING

Individual

MATERIALS

For each student:

✳ coin

✳ pencil and paper

✳ number cube (for Taking It Farther)

The Real Meal Deal

How many meals will students have to buy to collect all six toys from a fast-food chicken restaurant promotion? A simulation will give them a good idea.

⟳→ Directions

1. Distribute the reproducible and a number cube to each pair or group.

2. Explain to students that they will be testing to see how many meals they must buy to collect all six toys from a fast-food promotion. Before they begin, ask pairs or groups to predict how many meals they think it will probably take. Have them discuss the reasoning behind their answers.

3. Let students conduct the simulation on their own, rolling the number cube to represent each fast-food meal purchased. For each roll of the number cube, they make a tally mark next to the toy indicated by the number on the cube. For example, a 1 on the cube indicates Chaka the Chicken figure, and so on. Students keep rolling until they have collected every toy available. They repeat the simulation two more times and record the results in the columns.

4. When everyone is finished, allow time for students to compare answers to see how many rolls of the number cube it took to collect all the toys.

☆ Taking It Farther

Encourage students to look for real promotions where more than one item can be collected; for example, cereal boxes with promotional giveaways inside or soda bottle caps with special codes printed on them. Assuming there is an equal number of each item distributed, how long might it take to collect all of the items? Challenge students to design and conduct a simulation to find out.

✓ Assessing Skills

To make sure students understand the principle at work behind their simulations, have them do mathematical calculations to check the probability of collecting all six toys in the activity in only six tries. To do this, they should multiply: $\frac{1}{6} \times \frac{1}{6} \times \frac{1}{6} \times \frac{1}{6} \times \frac{1}{6} \times \frac{1}{6} = \frac{1}{46,656}$.

LEARNING OBJECTIVE

Students use a simulation experiment to see how long it might take to collect all the toys from a fast-food restaurant promotion.

GROUPING

Pairs or cooperative groups

MATERIALS

For each pair or group:
* *The Real Meal Deal* reproducible (p. 43)
* number cube

Name _____ Date _____

The Real Meal Deal

Jack's Chicken Shack is offering a special toy with each Chipper Chicken Meal.
There are six different toys, which are distributed into the meal boxes at random.
How many meals will you have to buy to collect all six toys?

Roll your number cube to see which toy you get. Match the number you roll with number of the toy. Make a tally mark in the Simulation A column next to that toy. Keep rolling the cube and making tally marks until you have collected each toy. Repeat the simulation for columns B and C.

CHIPPER MEAL TOY	SIMULATION A	SIMULATION B	SIMULATION C
1. Chaka the Chicken toy			
2. Chipper Chicken mobile			
3. Drumstick squeak toy			
4. Giblet Giggles joke book			
5. Peepers the Baby Chick wind-up toy			
6. Jack's Shack bouncing rubber egg			

What's Behind Door Number 2?

This problem was posed by Marilyn Vos Savant, a newspaper columnist who is reported by the *Guinness Book of World Records* to have the world's highest IQ.

Directions

1. Write the problem on the board.

   You're on a game show, and you're given a choice of three doors. Behind one door is a car; behind the other two are goats. You pick a door, say, door number 1. The host, who knows what's behind the doors, opens another door, say, number 3, which has a goat. He then asks you if you want to switch to door number 2. Would switching increase your chances of winning the car?

2. Have students discuss the question. Intuitively, most students will probably say that there is no point in switching doors, since there are now two doors, one with a car and one with a goat. The probabilities appear to be equal.

3. The intuitive answer is incorrect, says Vos Savant. Switching doors will increase your probability of winning the car. The reason? The host knows which door has the car. Use the following explanation:

   ✳ You pick door number 1. The probability that you have chosen the door with the car is $\frac{1}{3}$. The probability that the car is behind one of the other two doors is $\frac{2}{3}$.

   ✳ The host shows you what is behind door number 3. If you didn't pick the car, he will *never* show you the door with the car. Instead, he will always show you the door with the goat.

   ✳ The probability that the car is behind one of the doors you *didn't* pick is still $\frac{2}{3}$—but you know that one of them has a goat. If you switch to door number 2, you have a $\frac{2}{3}$ chance of winning the car. If you don't switch, the chance is $\frac{1}{3}$.

4. Many students will still be skeptical. Divide the class into pairs, with one student playing the part of the host and the other playing the contestant. Students model the problem 50 times, switching choices each time. Discuss the results as a class.

⭐ Taking It Farther

Let students play the game again, this time always keeping their original door choice instead of switching. What is the outcome?

✔ Assessing Skills

When they have completed the experiments, ask students to explain in their own words how the problem works.

LEARNING OBJECTIVE

Students conduct a simulation to understand a deceptively difficult probability problem.

GROUPING

Pairs

MATERIALS

For each pair:

✳ *What's Behind Door Number 2?* reproducible (p. 45)

✳ scissors

✳ 3 paper cups numbered 1, 2, 3

What's Behind Door Number 2?

Get ready to win a brand-new car . . . or maybe a brand-new goat!

Here's how to play the game:

1. Cut out the pictures of the goats and car below.

2. Decide who will be the game show host and who will be the contestant.

3. While the contestant looks away, the host puts one picture under each of three paper cups. The cups are the "doors." Only the host knows which door has the car.

4. The contestant picks a door. Then the host opens one of the other two doors—one that does not have a car behind it.

5. Now the contestant switches to the other unopened door.

Play the game 50 times, alternating roles. Each time, make a tally mark to show whether you won the car or a goat.

CONTESTANT'S NAME:	
WON CAR!	
WON GOAT!	

CONTESTANT'S NAME:	
WON CAR!	
WON GOAT!	

Elective Detective

Who is most likely to win the school election? Students use sampling to get a good estimate.

Directions

1. This activity is most appropriate for a time close to a school election. Begin by discussing polls conducted during city, state, and national elections. Have students noticed reports of these polls in the past? Discuss how polls are taken and how they work.

2. Now prepare for a small-scale sampling activity in your classroom. Write the names of the candidates for student council president on the board. Have each student write his or her vote for president on a slip of paper. Place all the slips of paper into a bag and call on a volunteer to randomly pick out 10 slips. Ask the volunteer to tally the votes on the board next to the candidates' names.

3. Ask students: *Based on these responses, how many total votes do you estimate the class has cast for each candidate? How would you make an estimate?* [One way is to use proportions to find an answer.] When you have discussed this, have another volunteer tally the remaining votes. How close were students' estimates?

4. Use all of the votes to have students predict how many votes will be cast for each candidate in the schoolwide election. How accurate might this estimate be? Discuss reasons why a student might be more likely to vote for one candidate instead of another, such as sharing a class with the candidate.

5. If possible, have students do a more randomized sample by randomly picking names of students in all the classes in your grade and interviewing them to find out their voting plans. Compare the resulting predictions to the predictions from the class poll. Then wait for the real election and see how close your predictions were!

Taking It Farther

Ask students to come up with other ideas for ways of taking a sample. Have them try their methods by surveying students about an issue such as voting age or a topic such as favorite television show.

Assessing Skills

Note whether students are able to use ratios and proportions correctly when trying to predict voting based on the sample.

LEARNING OBJECTIVE

Students conduct a survey to learn about sampling.

GROUPING

Whole class

MATERIALS

* pencil and paper
* paper bag
* list of all students in grade (if possible)

Picky, Picky

How many marbles of each color are in the bag? Students pick marbles and record their results to make predictions based on relative frequency.

➜ Directions

1. Divide the class into pairs and give each pair a paper bag.

2. Explain that one students will reach into the bag without looking, mix up the marbles, and pick one marble from the bag. The partner should will write the color on a sheet of paper.

3. The first student now returns the marble to the bag, mixes up the marbles again, and picks another marble. This process is repeated until it has been done 5 times, with the second student tallying the results.

4. At this point, students should try to guess how many marbles of each color are in the bag.

5. Students continue the activity until they have completed 10 draws from the bag and then guess the bag's contents again. Encourage them to revise their predictions as they continue to work.

6. Students repeat this process for 30 draws and 50 draws.

7. Finally, students look into the bag to check their predictions.

☆ Taking It Farther

Repeat the activity, this time placing 10 or more marbles in each bag and having students conduct more trials.

✓ Assessing Skills

Ask students:

✳ *Which predictions were most accurate?*

✳ *Why were they most accurate?*

✳ *What are some ways to check if a prediction will probably be correct or close to correct?*

LEARNING OBJECTIVE

Students find relative frequency and make predictions.

GROUPING

Pairs

MATERIALS

For each pair:

✳ brown paper bag

✳ 8 marbles of three different colors or other uniformly shaped items of different colors, such as colored paper clips

✳ pencil and paper

ADVANCE PREPARATION

Prepare brown paper bags by putting 8 colored marbles or other items in each bag. There should be three colors represented in each bag, with different proportions of colors.

Trick-or-Treat Numbers

At Halloween or any time of year, students will have a frightfully good time as they use candy to model the probability of independent events.

◉➤ Directions

1. Start by reviewing the probability of independent events. To find the probability of two independent events occurring, multiply the probabilities.

2. Bring out the bag of candy and explain its contents. Ask students:
   * *If you pick one candy at random, what is the sample space?* [0, 1, 2, 3, 4, 5, 6, 7, 8, 9]
   * *What is the probability you will pick a number greater than 5?* [$\frac{4}{10}$, or $\frac{2}{5}$]
   * *What is the probability you will pick a number less than 3?* [$\frac{3}{10}$]
   * *Say you pick one candy, return it to the bag, then pick another candy. What is the probability you will pick first a number greater than 5, and then a number less than 3?* [$\frac{2}{5} \cdot \frac{3}{10} = \frac{6}{50}$, or $\frac{3}{25}$]

3. Ask several more questions, such as, *What is the probability of drawing first an even number, and then and odd number?* or *What is the probability of drawing a number greater than 3 and then a number less than 7?* [$\frac{2}{5} \cdot \frac{1}{2} = \frac{2}{10}$, or $\frac{1}{5}$; $\frac{3}{5} \cdot \frac{7}{10} = \frac{21}{50}$]

4. For each question, have volunteers use the bag of candy to model the question several times. The class keeps track of the results of these experiments and discusses the outcomes.

★ Taking It Farther

Ask students: *How do these problems change if the candy is not returned to the bag after the first pick?* [They become problems of probability of dependent events, since the outcome of the first pick changes the sample space for the second pick.] Repeat the problems using this new method, and have students figure out the new probabilities. How do they compare?

✔ Assessing Skills

Note whether students understand that the probability for an event such as "greater than 7" is different from the probability for the event "7 or greater." Ask students to explain this in their own words.

LEARNING OBJECTIVE

Students find the probability of independent events and use models to investigate their answers.

GROUPING

Whole class

MATERIALS

* Halloween trick-or-treat bag or plastic trick-or-treat pumpkin
* 10 pieces of identically wrapped candy
* pencil or marker

ADVANCE PREPARATION

Use a pencil or marker to number the candies 0 through 9. Place them in the Halloween bag.

Game Show Showdown

Students take a new spin on the number 10 as they experiment with independent events.

➜ Directions

1. Describe the Go for a Spin! game to students. In the game, a wheel is divided into eight spaces labeled 1–8. Three players take turns spinning the wheel and can spin either once or twice. The object is to have the sum of the spins come as close to 10 as possible without exceeding 10.

2. To find the probability of getting two specific spins, you can multiply the probability of each spin. For example, to find the probability of spinning a 2 or lower on the first spin and a 2 or higher on the second spin, you'd multiply: $\frac{2}{8} \cdot \frac{7}{8} = \frac{14}{64} = \frac{7}{32}$.

3. Divide students into groups of three and let them play the game on their own. Students should develop their own strategies for winning. When all groups have played the game several times, let them share their strategies with the class. The class tries to decide which strategies work best and why.

★ Taking It Farther

Repeat the activity, but this time have students construct a large spinner with 20 spaces labeled in increments of 5¢ from 5¢ to $1.00. Challenge students to devise their own strategies for spinning close to $1.00 without going over.

Share with students that in 1993, a college math student named Steve Goodman used this information to develop a winning strategy for the showdown round of the TV game show *The Price Is Right*. For example, Goodman found that if the first player spins a 65¢ or lower on the first spin, he or she should spin again.

✔ Assessing Skills

Note whether students use their knowledge of the probability of independent events as they develop winning strategies.

LEARNING OBJECTIVE

Students develop strategies to win a game by finding the probability of independent events.

GROUPING

Cooperative groups of 3

MATERIALS

For each group:

* *Game Show Showdown* reproducible (p. 50)
* pencil and paper clip (to make the spinner)

Game Show Showdown

Hello! I'm Will S. Pinnen. Welcome to *Go for a Spin!*
How do you get close to 10 without going over? Try a
few different strategies and see. Use your knowledge
of probability of independent events to help you.
Good luck!

The object is to spin as close to 10 as possible, without going over.
Players take turns spinning. After a player's first turn, he or she can decide
to spin again, or decide to stay at one spin. Then it's the next player's turn.

SCORE SHEET

Game A	Player 1		Game D	Player 1	
	Player 2			Player 2	
	Player 3			Player 3	
Game B	Player 1		Game E	Player 1	
	Player 2			Player 2	
	Player 3			Player 3	
Game C	Player 1		Game F	Player 1	
	Player 2			Player 2	
	Player 3			Player 3	

Probability Scholastic Professional Books

Bon Voyage!

In this hands-on activity, students jet around the world with the probability of dependent events.

Directions

1. With dependent events, the outcome for one event depends on the outcome of a previous event. To find the probability, find the probability of the first event. Then use the new sample space to find the probability of the second event. If you want to find the probability that the two events will both happen, multiply the probabilities.

2. Ask volunteers to describe faraway places in the world they'd like to visit. Write the names of several destinations on the board. Direct each student to draw and cut out a picture of a plane ticket to one of the destinations. Each student may draw more than one ticket. Collect the tickets and make sure there are several tickets for each destination.

3. Place several tickets in a bag. Start with two destinations; for example, 6 tickets to China and 4 tickets to Morocco.

4. Have two volunteers say which of the two destination they'd like to visit. (Students may select the same destination.) Ask the class: *Picking at random from the bag, what is the probability that the first student will pick the destination of his or her choice?* Let the student pick and reveal her or his ticket. Now ask: *What is the probability that the second student will pick his or her choice? What is the probability that both students will pick the tickets of their choice?*

5. Continue modeling additional problems using more destinations and students. For each simulation, write the theoretical probability on the board before conducting the experiment.

☆ Taking It Farther

Have cooperative groups of students use the plane tickets to model their own problems. Each group should conduct its experiments repeatedly and share results with the class.

✓ Assessing Skills

Ask students: *If the probability that an event will happen is $\frac{1}{6}$, and you conduct the same experiment 6 times, is it certain that the event will happen?* [No. The results of a previous experiment do not affect the experiments that follow. The event could occur any number of times from 1 to 6, or not at all.]

LEARNING OBJECTIVE

Students conduct a simulation to test the theoretical probability of dependent events.

GROUPING

Whole class

MATERIALS

* drawing paper
* pencils
* crayons or colored markers
* scissors
* paper bag

Critter Cards

Which animal is on the card? As students guess and turn over each card, the probability they will guess right on the next card increases.

Directions

1. Review dependent events as necessary and distribute the reproducible. Each student cuts out the two sets of five animal pictures, placing one set in a row across his or her desk. Then he or she tapes or glues each of the matching pictures to the back of an index card, turns the cards facedown, and mixes them.

2. On a piece of graph paper, each student marks off five columns, labeling them Guess 1, Guess 2, and so on. Students then mark off 30 rows, labeling them Round 1, Round 2, Round 3, and so on to 30.

3. Begin by conducting Round 1 as a class, using one student's set of cards. Students guess which of the five animals is on the first card. Then turn over the card. If the guess was correct, students make a check under Guess 1 in the first row, Round 1. If the guess was incorrect, they make an X.

4. Now, place the card you picked over its matching animal cutout. For Guess 2, students guess which of the four remaining animals will appear on the second card. Repeat for all five cards.

5. To continue, students turn over the cards, mix them up, and make five more guesses. They should repeat this process 30 times.

Taking It Farther

Ask students:

* *What is the probability that you will guess correctly on the first guess?* [$\frac{1}{5}$]
* *What is the probability that you will guess correctly on the fifth guess?* [1]

Have students look at the proportion of correct guesses they made in each column. Ask: *How does the proportion compare with the theoretical probability? What is the probability that on one round of five guesses you will guess each animal correctly?* [$\frac{1}{5} \cdot \frac{1}{4} \cdot \frac{1}{3} \cdot \frac{1}{2} \cdot 1 = \frac{1}{120}$]

Assessing Skills

To see if students understand the term *dependent*, challenge them to think of events in their own lives that are dependent on other events.

LEARNING OBJECTIVE

Students conduct a hands-on activity to gain an understanding of the probability of dependent events.

GROUPING

Individual

MATERIALS

For each student:

* *Critter Cards* reproducible (p. 53)
* 5 index cards
* scissors
* tape or glue
* graph paper

Critter Cards

Cut out the animals. Tape or glue one set of animals to the back of index cards.

Card Tricks

An explanation for the probability of
mutually exclusive events is in the cards!

⟳→ Directions

1. Mutually exclusive events are two events that cannot happen at the
   same time. For example, when rolling a number cube, rolling an
   even number or rolling a 5 are mutually exclusive events. To find the
   probability of two mutually exclusive events occurring, add the
   probabilities. The probability of rolling either an even number or a
   5 is $\frac{1}{2} + \frac{1}{6}$, or $\frac{2}{3}$.

2. Display the deck of cards. Pose the probability questions below,
   asking students to describe the probability of the mutually exclusive
   events. Then have pairs model one of the problems at least 25 times
   to test the theoretical probability.

3. Note that in some problems, students must distinguish between
   mutually exclusive and non–mutually exclusive events. If an event is
   not mutually exclusive, it must be subtracted from the probability.
   For example, the probability of rolling an even number or a number
   less than 3 on a number cube is: $\frac{1}{2}$ (even numbers: 2, 4, 6) + $\frac{1}{3}$
   (numbers less than 3: 1, 2) – $\frac{1}{6}$ (2, which appeared in both sets) = $\frac{2}{3}$.

Probability questions:

* *What is the probability of picking an ace, king, queen, or jack?*
  [mutually exclusive: $\frac{4}{52} + \frac{4}{52} + \frac{4}{52} + \frac{4}{52} = \frac{16}{52}$, or $\frac{4}{13}$]

* *What is the probability of picking a club or a red face card (jack,
  queen, king, ace)?* [mutually exclusive: $\frac{13}{52} + \frac{8}{52} = \frac{21}{52}$]

* *What is the probability of picking a number card greater than 7 or a
  jack?* [mutually exclusive: $\frac{12}{52} + \frac{4}{52} = \frac{16}{52}$, or $\frac{4}{13}$]

* *What is the probability of picking a number card less than 8 or a
  diamond?* [non–mutually exclusive: $\frac{28}{52} + \frac{13}{52} - \frac{7}{52} = \frac{34}{52}$, or $\frac{17}{26}$]

* *What is the probability of picking an odd number card or a black
  card?* [non–mutually exclusive: $\frac{16}{52} + \frac{26}{52} - \frac{8}{52} = \frac{34}{52}$, or $\frac{17}{26}$]

✪ Taking It Farther

Have students write a list of mutually exclusive events from their own
lives. Hold a class contest to see who can think of the most examples.

✓ Assessing Skills

Observe whether students are able to distinguish between mutually
exclusive and non–mutually exclusive events.

Mutually Exclusive Events

LEARNING OBJECTIVE
Students find the probability of
mutually exclusive events.

GROUPING
Whole class/pairs

MATERIALS
* deck of playing cards
* scrap paper and pencil

Wheel of Fortune

Students step right up and learn how odds come into play at a county fair!

⟳→ Directions

1. Duplicate the reproducible.

2. Discuss the concept of *odds* with students. When have they heard the term before? Students may not realize that the odds of an event are different from the probability of an event. On the board, write this definition of odds as related to winning a game:

$$\text{Odds of winning} = \frac{\text{Number of winning outcomes}}{\text{Number of losing outcomes}}$$

The *probability* of winning compares the following:

$$\text{Probability of winning} = \frac{\text{Number of winning outcomes}}{\text{Number of all possible outcomes}}$$

3. Distribute the reproducible. Have students color in alternating sections of the spinner in light blue and yellow, so that half of the sections are yellow and half are blue. Make sure that the colors do not obscure the spinner numbers.

4. Students complete the reproducible on their own.

☆ Taking It Farther

Let students go back to questions 2–5 on the reproducible and determine the probability of the outcomes.

✓ Assessing Skills

Observe whether students compute the odds of winning or losing, as opposed to the probability.

LEARNING OBJECTIVE

Students learn about odds and practice applying the concept in a game.

GROUPING

Individual

MATERIALS

* *Wheel of Fortune* reproducible (p. 56)

* light blue and yellow crayons or colored pencils

* pencil and paper clip (to make the spinner)

ANSWERS

1a. 1

1b. 16

1c. 24

2a. $\frac{1}{47}$

2b. $\frac{16}{32}$, or $\frac{1}{2}$

2c. $\frac{24}{24}$, or $\frac{1}{1}$, or the odds are even.

3a–5b. Answers will vary. Check students' results.

Wheel of Fortune

Note: Before playing the game, color in alternating sections of the spinner with light blue and yellow crayons or pencils.

Poor Louie! Here at the County Fair, he's placed ten bets in a row on 12—his lucky number. But the number hasn't been so lucky for Louie today. He's lost every time!

Help Louie understand why by using the Wheel of Fortune to answer the questions about odds and test your answers. Remember, in this game you can bet that the spinner will land on any number, shape, or color. Good luck!

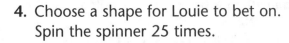

1. How many chances does Louie have to win if he bets on

   a. one number?   _____

   b. one shape?   _____

   c. one color?   _____

2. What are Louie's *odds* of winning if he bets on

   a. one number?   _____

   b. one shape?   _____

   c. one color?   _____

3. Choose a number for Louie to bet on. Spin the spinner 25 times.

   a. How many times did Louie win?

   _____

   b. How many times did Louie lose?

   _____

4. Choose a shape for Louie to bet on. Spin the spinner 25 times.

   a. How many times did Louie win?

   _____

   b. How many times did Louie lose?

   _____

5. Choose a color for Louie to bet on. Spin the spinner 25 times.

   a. How many times did Louie win?

   _____

   b. How many times did Louie lose?

   _____

Probability Scholastic Professional Books

Above Average

Mean, median, or mode? Student statisticians decide!

◌→ Directions

1. Usually, an average refers to a mean. But statisticians sometimes use the median or mode of a group to make the most accurate predictions. It all depends on what is appropriate for the data collected and what type of prediction you want to make.

2. For example, consider the shoe sizes of a group of people.
   ✳ The *mean* is the sum of all the sizes divided by the number of people in the group.
   ✳ The *median* is the number exactly in the middle of the group—an equal number of people have a smaller shoe size and an equal number have a larger shoe size. If one or two people have a much smaller or a larger size than the rest of the group, the median may give you a better idea of the size of the group's feet.
   ✳ The *mode* is the shoe size that occurs most often in the group. To find out which size you would be most likely to get if you picked a person at random, you would use the mode. A mode can also be used as the average when many of the measurements in a group are the same or similar.

3. Have students provide some or all of the following information:
   ✳ how far the student travels to get to school each day
   ✳ shoe size or length of foot in centimeters
   ✳ height in inches

4. Tally the results and let students use the information to make predictions about the rest of the students in their grade. For example, what would be the most likely height for a randomly selected student? Discuss whether mean, median, or mode is the most effective way to represent each result.

☆ Taking It Farther

Ask students to come up with fictional scenarios in which it would be most useful to make predictions based on a mean, median or mode.

✓ Assessing Skills

Observe how students differentiate among *mean*, *median*, *mode*, and *average*. Can they clearly explain their reasoning about which to use?

LEARNING OBJECTIVE

Students explore why statisticians might choose to use mean, median, or mode when making different kinds of predictions.

GROUPING

Whole class

MATERIALS

✳ rulers and other measuring tools (optional)

Rain, Rain, Go Away

What does "a 40 percent chance of rain" really mean? Students get an idea as they keep track of weather forecasts and compare them to the actual weather.

➤ Directions

1. Discuss weather forecasting with the class. Ask students if they have heard forecasts that included a percent chance of showers, snow, and so on. Can they explain what the forecast meant? How would they write "a 40 percent chance" as a fraction?

2. Each day for several weeks, have students check the same source (a cable TV weather channel, a newspaper, or a radio station) for the weather forecast for the following day.

3. On days when a specific weather condition such as snow or rain is forecast, ask students to record the percent chance of the condition in their notebooks.

4. The next day, students record the actual weather that occurred beside the forecast.

5. After two or more weeks, let students read through the notebooks to analyze the results. Do the predictions seem accurate? What are some reasons why the forecasts might or might not have been accurate?

★ Taking It Farther

On a day when a percent chance of rain, snow, or another weather condition is forecast, have students use fractions to create a circle graph spinner. The percent chance of rain or snow should be colored in and the rest of the graph left blank. If there was the same chance of rain or snow for 15 days in a row, on how many of those days might it actually rain or snow? Have students spin repeatedly to get an idea of what could happen.

✓ Assessing Skills

Ask students: *If there is a 40 percent chance of rain on one day, and a 60 percent chance of rain on a second day, what is the probability that it will rain on both days?* [Since these are independent events, multiply $\frac{2}{5} \bullet \frac{3}{5} = \frac{6}{25}$.]

LEARNING OBJECTIVE

Students follow weather reports to gain an understanding of probability and chance.

GROUPING

Whole class

MATERIALS

✴ daily weather reports from newspapers, radio, or television

✴ notebook for recording forecasts and weather observations

✴ scissors

✴ cardboard

✴ colored pencils or markers

✴ pencil and paper clip (to make the spinner)

Pick a Player

Students use probability to experience what it's like to be a team owner picking players in the NBA draft lottery.

⟿ Directions

1. Divide the class into groups of four. Distribute a copy of the reproducible to each group, along with the rest of the materials.

2. Read the introduction to the reproducible and discuss the NBA draft lottery. Then explain that students will be simulating the lottery in this activity.

3. Have volunteers write the names of their favorite basketball players on the board. In this simulation, the class pretends that these are new players who are eligible for the NBA draft. Students may cite male and female players, as well as professional and nonprofessional players (including schoolmates).

4. Each student will represent several teams, since the lottery is held for thirteen teams. For instance, three students may each represent three teams, and the remaining student can represent four teams. Each group member writes his or her name in the Draft Lottery Chart under Team Owner for each team he or she "owns."

5. Each group writes the numbers from 1 to 1,000 on separate strips of paper and places them in a bag. They shake the bag to mix the numbers, and a student draws a number. The owner who has that number in his or her range of Team Draft Numbers picks the first player from the list on the board. A second and third number are drawn, and the corresponding owners pick players.

6. When the teams are complete, students return the numbers to the bag and repeat the simulation several times with additional copies of the reproducible. What conclusions can they draw?

★ Taking It Farther

Ask students to write a journal entry to say whether they feel that the NBA draft lottery is fair, why or why not, and whether their opinions were changed by the simulation.

✓ Assessing Skills

As students conduct the simulation, ask them to describe the probability that each of their teams will get the next number. Note whether students remember to account for a reduced sample space after numbers have been drawn.

LEARNING OBJECTIVE

Students discover how probability of dependent events affects the NBA draft lottery.

GROUPING

Cooperative groups of 4

MATERIALS

For each group:
* *Pick a Player* reproducible (p. 60)
* paper bag
* scissors
* paper
* markers

59

Pick a Player

Each year, the 13 National Basketball Association teams with the worst records participate in the NBA draft lottery. Each team is assigned a group of number sets. The worse a team's win-loss record is, the more number sets that team gets. A number set is randomly drawn, and the owner of the team who has that set gets to draft, or pick, the first new player of the year. Would you like this system if you were a team owner? Try it and see!

Decide who "owns" Team 1, Team 2, and so on. Take turns drawing numbers from the paper bag. These are your Team Draft Numbers. If the draft number belongs to the team you own, pick a player from the list.

DRAFT LOTTERY CHART

	TEAM OWNER	TEAM DRAFT NUMBERS	PLAYER(S) PICKED
Team 1		1–250	
Team 2		251–450	
Team 3		451–607	
Team 4		608–727	
Team 5		728–804	
Team 6		805–880	
Team 7		881–924	
Team 8		925–953	
Team 9		954–971	
Team 10		972–982	
Team 11		983–989	
Team 12		990–995	
Team 13		996–1,000	

Do I Have Problems!

Use these quick skill builders as a class warm-up, as a time filler for students who finish a test early, or just for a fun break from the textbook!

7-11 BIRTHDAY

Who wants a happy birthday Slurpee? You pick a person at random out of a crowd. What is the probability that the person's birthday falls on either the seventh or eleventh day of a month? Write the answer as a fraction, a decimal rounded to the nearest thousandth, and a percent rounded to the nearest tenth. When answering, disregard February 29, which occurs once every four years at leap year. [$\frac{24}{365}$, 0.066, or 6.6%]

SPIN CITY

Draw the spinner below on the board. Then ask students to solve this problem: *The spinner is spun three times. How many combinations of spins are possible?* (If students need a hint, suggest that they use the Fundamental Counting Principle to find an answer.) [6 × 6 × 6, or 216 combinations are possible.]

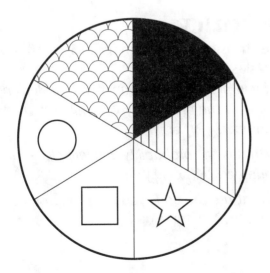

COOKIE CONUNDRUM

Suppose you are making a batch of six cookies. You drop ten chocolate chips into the batter and mix it up. If the chips are randomly distributed among the cookies, how likely do you think it is that you will get a cookie with three or more chocolate chips? Have students come up with a theory and test it by rolling a number cube ten times. Each time a number is rolled, it represents one chocolate chip going to a cookie. For example, rolling a 2 means that cookie number 2 gets one chip.

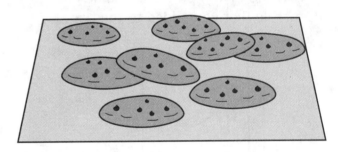

WHAT'S THE NUMBER?

You're having trouble remembering a friend's phone number. You know the first three digits, but you can remember only that the last four digits consist of a 2, 4, 7, and 8 in some order. How many ways could you arrange the four numbers to finish the phone number? (Hint: Use a tree diagram to help you find out.) [4 × 3 × 2 × 1 = 24] What if the last four numbers were 2, 4, 4, and 7? How many different arrangements could you make then? [12]

IN THE BAG

Write the following brainteaser on the board. Before revealing and explaining the answer, have students discuss their ideas.

You are holding a bag containing one marble. You know the marble is either red or black. You take a red marble, add it to the bag, and shake the bag. Then you reach in without looking and pick out a marble. It is red. What is the probability that the marble left in the bag is also red?

[The probability that the marble left inside the bag is red is $\frac{2}{3}$. That's because there are three possible scenarios in which you might pick a red marble from the bag. In two of those three scenarios, both marbles are red. The scenarios are:

1. The original marble was black. You reached in and picked the same red marble that you added to the bag.

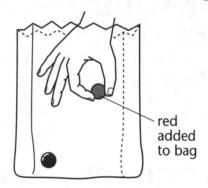

red added to bag

2. The original marble was red. You reached in and picked out the original marble.

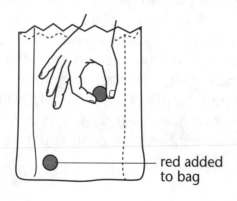

red added to bag

3. The original marble was red. You reached in and picked the same red marble that you added to the bag.

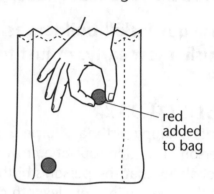

red added to bag

You can simulate the problem with the class. You'll need a small bag and three marbles, balls, or plastic chips—two of one color, such as red, and one of another color, such as black. Place one red or black marble in the bag so that the class cannot see. Hand the bag to a student, have the student place a red marble in the bag, shake the bag, and pick out a marble. If the marble is red, have the student guess what color the other marble is. Ask students to model this experiment repeatedly and keep track of the results. After each trial, someone should change, or pretend to change, the marble inside the bag.]

HAT TRICK

Place strips of paper numbered 1 to 10 in a hat. Ask students:

✳ *If you pick a strip of paper without looking, what is the probability you will pick an even number?* [$\frac{5}{10}$, or $\frac{1}{2}$]

✳ *What is the probability you will pick a prime number?* [$\frac{4}{10}$, or $\frac{2}{5}$]

✳ *What are the odds you will pick a prime number?* [$\frac{4}{6}$, or $\frac{2}{3}$]

 Name _____   Date _____

In My Opinion

The activity _____ was:

Easy   Hard

because:

_____

_____

_____

_____

My work on this activity was:

poor   fair   good   excellent

because:

_____

_____

_____

I used the following math strategy or strategies:

_____   _____

_____   _____

_____   _____

I would share this tip with someone who is about to do this activity:

_____

_____

_____

Activity _____ Date _____

Student						
UNDERSTANDING						
Identifies the problem or task.						
Understands the math concept.						
SOLVING						
Develops and carries out a plan.						
Uses strategies, models, and tools effectively.						
DECIDING						
Is able to convey reasoning behind decision making.						
Understands why approach did or didn't work.						
LEARNING						
Comments on solution.						
Connects solution to other math or real-world applications.						
Makes general rule about solution or extends it to a more complicated problem.						
COMMUNICATING						
Understands and uses mathematical language effectively.						
COLLABORATING						
Participates by sharing ideas with partner or group members.						
Listens to partner or other group members.						
ACCOMPLISHING						
Shows progress in problem solving.						
Undertakes difficult tasks and perseveres in solving them.						
Is confident of mathematical abilities.						

SCORING RUBRIC

3	2	1
Fully accomplishes the task.	Partially accomplishes the task.	Does not accomplish the task.
Shows full understanding of key mathematical idea(s).	Shows partial understanding of key mathematical idea(s).	Shows little or no grasp of key mathematical idea(s).
Communicates thinking clearly using oral explanation or written, symbolic, or visual means.	Oral or written explanation partially communicates thinking but is incomplete, misdirected, or not clearly presented.	Recorded work or oral explanation is fragmented and not understandable.